SIMULATION THEORY

MANAV SHARMA

Copyright © Manav Sharma
All Rights Reserved.

ISBN 978-1-63745-951-5

This book has been published with all efforts taken to make the material error-free after the consent of the author. However, the author and the publisher do not assume and hereby disclaim any liability to any party for any loss, damage, or disruption caused by errors or omissions, whether such errors or omissions result from negligence, accident, or any other cause.

While every effort has been made to avoid any mistake or omission, this publication is being sold on the condition and understanding that neither the author nor the publishers or printers would be liable in any manner to any person by reason of any mistake or omission in this publication or for any action taken or omitted to be taken or advice rendered or accepted on the basis of this work. For any defect in printing or binding the publishers will be liable only to replace the defective copy by another copy of this work then available.

Contents

Acknowledgements	v
Preface	vii
Prologue	ix
1. Simulation Theory	1
2. God	3
3. Us In The Simulation	4
4. Time And Existence	5
Conclusion	7

Acknowledgements

I give special credit to Dr Rizwan Virk from Massachusetts University and the founder of PlayLabs at MIT. He is the author of The Simulation Hypothesis and is one of the few responsible for the entirety of this book.

Preface

Everything that we currently have in this vast world of ours baffles me. It all seems too good to be true, be it a terrible catastrophe or a remarkable achievement. The past, present, future- is it all real? Or is this all just a simulation to keep us in line? To remind us of just how mortal we are?

Prologue

This is a book discussing Simulation Theory, a published hypothesis by Dr. Rizwan Virk of MIT, which opens up to the idea of us collectively living in a fake reality which is not acknowledged by us in our everyday life. It explores the possibility of everything we know to be a fake reality in which we are set. The book also discusses concepts of time, and existence itself.

CHAPTER I

Simulation Theory

Do we live in a simulation? There is no way of proving it, but there is a way of justifying it.

I have not been made aware of anything yet. 'Anything' does not constitute the mannerisms that I have experienced in my time so far, and also the ones I continue to experience. It constitutes that of what was, is, and will happen. I am merely a microscopic speck in the universe, awaiting my destiny if something like it even exists. If this is a simulation, after all, I am not able to control anything. I am a specimen of this experiment, and this series of events have already taken place. The 'future', entirely, has not been looked at, nor will it ever be told of as to my understanding. To my understanding, the 'future' is something to be predicted; it is of pure 'luck' to come across a prediction that is completely true and that the events taking place in the prediction happen in the same order as predicted. The prediction is made according to what is observed. These observations if changed, change the predictability of the future, making it utterly false, or completely identical. The point trying to be conveyed is that if the pieces of our life shift around in a manner which we think of as being completely natural order, can change the way we perceive the future. Having a look at the big picture, I am nearly definitive of me being in a simulation. If it is just I, then everything around me does exist fundamentally but exists in a virtual platform of someone's design. On the other hand, it would be jarring to suddenly wander off this mainstream yet simulated reality; I have suited to 'here' as my existence. This existence of mine is not only catered to me but the countless virtual folk around me; as exhibited by the person or people running this simulation.

CHAPTER II

God

It is not possible for a God to be running the show, at least in my terms; God to me is a mysterious construct of the mind who has been imagined as someone who can decide for you; whatever the decision may be drafted from. This typing may also just be a simulation for me, or the reader, designated for a function no one is aware of. There is a possibility of everyone experiencing the same simulation, which makes this a game of sorts. People and events are moved around like chess pieces if chess even exists. Everything is a construct, and nothing is a reality in my virtual reality which I thought of to be real or is expected to be real. Any moment of history could not even exist; it could be put there as encouragement for creating and/or manipulating the present/future. Our losses and victories act as catalysts of hope; they did it so we can too; This simulation could also be created by an advanced species; ourselves by theory, to find and test out the best possible reality for them. If this is run by an entire species, then something in their 'real' reality is not right; that is the reason for these simulations. This entire typing is hypothetical and based on the pure theorem. Is this all just a creative visualisation?

CHAPTER III

Us in The Simulation

Taking me into perspective as an equal part of this simulation, to explain what I am trying to convey, I can justify some aspects. I am not obligated to do this, although I am unaware as to why I choose to do so. When you see yourself from your own eyes, you are only able to observe your hands, torso, legs, and everything else available before you. You see hints of your face, but that is all on the exterior. What is not seen is your internals, which is revealed to you once you look inside. You expect them to be present in a generalized order about which most of us know about. This expectation leads to a sense of belief, as I think. My words cannot or shall not be definitive, since I think of myself to be an individual among other individuals living in a simulated reality with a human characterization.

What I can say for sure is that opinions change regardless of you living in a simulation or not. You may comfort or reassure yourself every day that you are just' doing fine'. The reality is that you are not doing fine, but are wasting your time. This sounds incredibly grim, but it is the truth.
We in the simulation are somewhat like characters from a game; games have been advancing in better and different ways through the ages. This is just another advancement in a game, one that is so advanced that we have sunken in deep enough to not realize that everything around us has been virtually created by an advanced race, the race likely being us ourselves.

CHAPTER IV

Time and Existence

Take a step back to realise that you nor your existence actually matters and that even if it does, there is no way of knowing it. If there is a way, and if that way is accomplished, it defeats the purpose of you spending your non-precious time searching for it because it will never be a reality, nor should it be. As I continue to type this, I have realized that this has turned into something more than that about a simulation. This is foolish trickery, as I'm led to believe.

When we think of all that has and will happen, we overlook the fact that it is either happened or is going to happen. There is no conformation in this, or anything, other than shreds of evidence found for something that has happened, and there are only assumptions or speculations, and ever-changing information obtained from preliminary sources such as the present. What I say is that a simulation consists of the so-called past, present, and the future at the same time, with all three of them changing at any given moment of existence. We are not in control of changing anything; it changes itself or is changed by someone about whom we will never know.

Time itself goes by without us knowing it. We characterize it through a number or a letter, which gives it a representative value, thus making us understand it. This is how we perceive time. Existence and Time go hand in hand; existence is counted on as moving forward through time, as time is characterised through letters and numbers. We can be in a virtually simulated environment, perhaps one of our design.

Conclusion

This concludes my thoughts on the idea of us as a society and species living in a virtually constructed simulation and the fact that all of our days might just be a fake reality. I conclude by saying that this is only a thought, a possibility, which is a strong one. Living in a simulation or not, we as one species will continue to strive in any possibility that strikes us. We will continue to prosper and will continue to fall in our ways. Everything will continue, evolving as time goes by, but remaining the same. Thank you, reader.

www.ingramcontent.com/pod-product-compliance
Lightning Source LLC
Chambersburg PA
CBHW020716180526
45163CB00008B/3122